WHALE SONG

WHALE SONG

Jay Amberg

Third Edition, Revised 2020; First published 2009   ISBN 13: 978-0-9708416-5-0

AMIKA PRESS   466 Central AVE #23 Northfield IL 60093   847 920 8084
info@amikapress.com   Available for purchase on amikapress.com

100% of author royalties will be donated to The Nature Conservancy.

Edited by John Manos. Designed and typeset by Sarah Koz. Titles set in Perpetua, body in Joanna. Both designed by Eric Gill in 1925 & 1930 respectively. Ornaments set in Decoration Pi 1 by Adobe, 2007. Thanks to Nathan Matteson.

To my father, Thomas H. Amberg
& for his brothers and sisters—
Agnes, Bill, Bud, Dinny, Gil,
Ibby, John, Katie, Mary and Peg

I came into this world flukes first. Light broke around me, vast and blue. Stroking me, my mother nudged me toward air. My first breath was tangy, my lungs filling and every cell welling as I breached in those Pacific waters.

My flukes unfolded. Two of my aunts prodded me along. Clumsily, I swam in the swells. I needed my mother's milk as much as that August air, and so I dove beneath her. As she slowed, I took her nipple. Her sweet rich milk streamed down my throat. My aunts and cousins sang to me. All about me, the sea and sky clicked.

❦

This is my song, a hymn of life—my life and yours. My song is at once beautiful and painful. At times a ballad and an anthem. By turns a psalm and a lampoon—as we would expect from life. Not often an idyll, and never a lullaby. Not a dirge either, for I am not submerged in the inevitable losses of our lives. But a chant, a canticle, a litany.

I am, by nature, gentle. Even shy, at least around those outside my family and friends, my clan. And I hope you will remember this because some of what I sing is neither gentle nor bashful. When, at times, my song sounds harsh, it is that so much is at risk—our survival, certainly, and yours, as well. I do know that we are bound, as sure as a harpoon is to its line. So know that even when my words are sharp, when my barbs strike deep, I am, at the far end of the line, a gentle soul.

You have not yet understood or heeded our songs. Perhaps, you never will. But I will do my best to have you hear me before it is too late. Before all of the songs, ours and yours, sink into time's depth and darkness. Listen. My life depends on it. Your life depends on it.

❦

I weighed only a ton when I was born, but now I am enormous.

My mother's milk was high in protein and minerals and, yes, fat—and I grew prodigiously. Now, as I approach my sixtieth birthday, I am just under sixty feet long and weigh fifty-nine tons. Indeed, now I *eat* a ton on a good day, and my blubber remains thick. My four-hundred-pound heart pumps five gallons of blood at a beat. Despite my gargantuan size, I am fast. Normally, I travel at four to seven knots; I dive at three. But I can swim at twenty knots when I need to. I'm not bragging. I just need you to know who I am.

My eyes look small because of my bulk, but they are two-and-half inches in diameter. Absolute size controls functioning in eyes—as it does, to a certain extent, in brains. My eyes really are large, and I see well—except, of course, that my huge head causes quite a blind spot. In fact, my head is one-third of my body, about twenty feet and twenty tons: I am the largest of the Toothed Cetaceans. But Blue Baleens are far bigger— the largest animals who have ever lived in this world. So my size is relative, titanic to you but not to a Blue.

⋙⋘

I don't really look like anything else in this world. You would, I suspect, think me ugly. And, I suppose I am. My head is squarish —*blockish*, you might say. It contains the largest brain ever as well as vast amounts of spermaceti, the oil for which you slaughtered my ancestors. My blowhole is a slit on the left near the front of my head. My blow is bushy, forward, and skewed to the left.

My eyes bulge a bit. I have no visible ears, but I hear far better in water than you. Sound travels through my inner ears, my jaw, and my spermaceti cavity. I have no hair, as most other mammals do, but my blubber keeps me warm and, because it is lighter than water, buoyant. I can open my jaws ninety degrees. My lower jaw is long and narrow and underslung. My teeth—conical, thick, and heavy—fit into slots in my wide upper jaw, which is toothless. My throat is also wide, broad enough, in fact, for you to pass down it.

My skin is creased, wrinkled, unlike the smooth skin of other Cetaceans. In the millions of generations since my forebears returned to the water, my arms evolved into short fins with rounded tips. My back has no dorsal fin, just a hump and a series of knobs running toward my tail. My torso is extremely muscular (I have, after all, to propel fifty-nine tons at speed), and I have no vestigial legs at all. My intestines contain ambergris, for which you also massacred my fore-fathers. It is nothing more than an intestinal by-product, but you still value it beyond measure.

My broad triangular flukes are horizontal—enough for you

to know that I am no fish. The trailing edges of my flukes are pretty well frayed. *Scalloped* by life. I have been around a long time, roved the world, experienced much that living offers us all. I realize that my longevity isn't exceptional for a large-brained mammal, but in these last sixty years the changes in our world have been epochal, even apocalyptic. Of these I will sing.

Whenever one of you has spotted me, you have stalked me. You no longer have murder in your hearts, but you still cannot leave me be. I suppose it is my color. I am large, of course, but not much larger than some of my cousins or, had he lived, my brother. My whiteness gets your attention.

Your stalking, understandably, bothers me. My breath quickens, and I don't sing as much. I spend more time on the surface, and I find myself changing course more often than I would like. It's not just your constant noise, though that confuses all of us. Your presence out here doesn't make sense, given all you should be doing on land at this moment.

This pallor of mine occurs every four generations or so in my family. It is not a curse. No one has shunned me, and mating has been anything but a problem. And yet my hue sets me apart. Always has. But it doesn't change who I am. It just is. My color has never mattered to me or to my family and friends. Why should it?

⁓

My spermaceti organ protrudes from my head, making my snout stout. You might think *bloated*, even *engorged*. It appears swollen in large males, and I am an especially large male. It runs from under my blowhole to a crease at the back of my head—in my case, more than sixteen feet, with an air sac fore and aft. All in all, a good sized Pilot Whale could swim inside my head.

Using my spermaceti organ, I can produce sounds in pulses that allow me to echolocate and communicate over long distances. I can, whenever I need to, expand my sensory field for many miles. I can also produce sounds intense enough to stun prey and frighten competitors. And, well, females like my organ—both for its size and the intensity of my clicks and calls, codas and chants. You see, for us, acoustic size matters. I sound even larger than I am, giving me certain advantages in attracting females and daunting other males. Then too, my head provides me, when necessary, with an impressive battering ram.

You have always coveted the oil inside my organ. You first slaughtered my forebears for this fluid I use to navigate and communicate. You could have learned a lot from studying our spermaceti organs, these innate systems that do so much of what you need to understand about navigating and communicating. But instead you massacred us. You demonized us and butchered us, sending us to the edge of extinction, but mostly you misunderstood us. You called us *sperm whales* because this oil, an absolute necessity to us and merely a commodity to you, looked like your sperm.

I dive deep. Deeper and longer than any other animal. The greatest breath-holding diver who has ever lived. Often I dive to more than 3,600 feet for an hour at a time. But I have sounded for twice that long to far more than a mile. Boasting again? I hope not. Just letting you know what it is that I do.

I throw my flukes in the air and plunge vertically. My flukes beat hard, but my heart rate drops as I descend. My lungs are extremely efficient, and my muscles store oxygen well. I dive, as I said, at about three knots. My motto is *Do not rush; do not rest.* If I go too fast, I burn too much oxygen; too slowly, and I burn time. At a depth of 800 feet, my lungs and jointed ribcage collapse. Light diminishes, and I start to click. My echolocation is so sophisticated that I really can *see* in sound as the world around me sinks into darkness.

By the time I reach 1,600 feet, it's absolutely dark. The world is cold. Time alters, slows, then vanishes. I can hear distant sounds—storms and earthquakes and the incessant buzzing of your ships. Sometimes, you emit even more discordant sounds, sonar signals and seismic pulses.

At depth, I hunt for squid and octopus; deep-dwelling sharks are good sport, too. Below 3,200 feet, my lungs are flat, but air in my nasal passages still circulates so that I can reflect sounds and create clicks. I flow, lit in an acoustical world. Like you, I am composed mostly of water. And when I go especially deep, I become water. The world beyond, air and sky and sun and stars, evaporates in the aquatic moment.

And here I meet Architeuthis, the giant squid, my most worthy foe—and my favorite food. Though I am one hundred times heavier than Architeuthis, the outcomes of our battles are never foregone. The greatest invertebrate in the ocean, he is superbly fit for life in the deep, with a large brain and gigantic unblinking eyes that provide him with sharp sight in what appears to be total darkness. His funnel propels him forward and backward; ink from his sac provides a concealing cloud. The ringed suckers on his eight arms and two long tentacles are toothy, and his hooked beak is sharp. Indeed, with him I can never be certain that I will return from the hunt's dark depths.

⁂

My life is a cycle, the pace slow. I was in my mother's womb fifteen months. I took milk for three years. I was twenty-nine before I was a father. I have been roving, North and South, East and West for more than fifty years. The ocean, my home, cycles with the tides, the day, the moon, the seasons. Though the current runs north and west around the tip of Alaska, it flows south along the rest of the continent. Near the equator, it heads west toward Asia where it turns again and circles back, forming the mammoth North Pacific Gyre. Even our home, the world itself, cycles through the year and through the ages. My pace within these cycles provides me with time to think. My thoughts cycle, too, between good and evil, comedy and tragedy, hope and despair.

Your thoughts would cycle, as well—if you took the time. But you are too busy. And your busyness, your business, hurtles you toward the apocalypse. There will be flood and fire, strife and starvation, carnage and catastrophe. Your home, our world, will howl.

And yet, and yet my thoughts cycle. Our destruction isn't

predestined. At this moment, the light dances on the water, the sparse clouds form runes, and the current whispers to me, even as the melting ice moans and the calving bergs groan. Ice has always sung its own sad song, but it is transmuting now —fast becoming a requiem. Here and in the Southern Ocean.

Life has scarred me from head to flukes. It can't be helped, of course. Life does it to us all. The front and top of my head are thoroughly cross-hatched. Tattooed by the beaks and suckers of Architeuthis. But, in truth, my head is scored as much by the teeth of other Cetaceans.

We males fight. We compete for the right to mate. None of us mates every year, and some of us never become fathers. Our battles are real, not ritualistic. When I was younger, I fought hard, attacked with my whole being. I gnashed my teeth and flailed my flukes. I broke jaws, but I never killed another Cetacean. Never.

Even though I didn't lose a battle for a decade and a half, I only fathered five offspring. Our reproductive rate is lower than that of any other animal. Three of my daughters survived to adulthood, and two are now themselves mothers. One has already lost a baby, an event from which she has never quite recovered. You see, females only bear a child every five or six years. Twins are rare. The young die far too often, and mothers do not easily get over their deaths. My daughter carried her son in her mouth for a week after he died.

⁂

The sun is heading south. Fall is here, and I am leaving. It is time. Time to descend from the higher latitudes as I sing to you. Time to return to the Galapagos, to the sea of my birth where this song began for me.

Winter is coming, but even here in the Gulf of Alaska it is not what it once was. Fall arrives a bit later now, and spring earlier. I can feel the difference in the water. It is warmer, perhaps even a full degree. The current is changing, too—not a lot yet but enough so that we who live here can't help but notice the shift. And all life as we know it will turn with the current. I hope you are finally beginning to comprehend how much the currents affect us all. How much the currents influence even your life.

I have put Polaris behind my flukes. I am again pelagic, the seafarer roving one last time. Everything wheels through the night sky around the polestar. The wheeling is beautiful, the night sky cycling as the year does. And our lives. But the

North Star, fixed, is guide and constant companion to a solitary wanderer. Though it is immensely distant, it seems close,
its light touching my back whenever I breathe. Constancy does
that. Gives us the feeling, the illusion, if you will, of closeness.
And I understand that even Polaris is slowly shifting, that one
night it, too, will wheel, no longer due north, the marker of
the geomagnetic pole. But I will be long gone. So may we all.
My thoughts are concerned with the current spinning, the
changes that have accelerated in my lifetime.

I don't know if I will ever return to this gulf. I will miss the
flash of light on the distant ice. I will miss the sun shimmer-
ing on the water. I will especially miss the Aurora Borealis, the
light wavering in darkness, pulsing to the beat of the world, its
song calling out to the stars, to the universe of which we are
such a small but significant part. I'm not a fatalist, but this may
be my last journey, my last song. The world is changing fast,
and I am no longer young enough to ignore what is occurring.

⚮

I will not, however, miss your presence in this gulf. The changes
you wreak are radical. I was here, you must remember, when
your wayward tanker spewed your crude oil into our waters. I
was just returning in the spring of my forty-second year when

the tanker tore out its belly and disgorged 44,000 tons of black sludge. I was able to stay far enough from shore to avoid this spreading black plague, but everyone else, from fish to fowl, was infected. Many small marine mammals, particularly seals and otters, perished. The web of life was altered here—perhaps not permanently, but certainly for my lifetime and yours.

And that Black Death almost two decades ago was relatively small—only 1,500 miles of shoreline were damaged. There have been many worse spills, including ten of 95,000 tons or more and four of more than 200,000 tons. But this plague sticks in my mind because it happened in my home—and the recovery will never fully occur in my life. Ironically, the disaster has only driven more of you here with more of your noise and more of your debris discharged into the gulf. I take it all personally because I live here. But you have to understand that the coming catastrophe will be personal for *all* of us, whoever we are and wherever we live.

⤫

As I said, my brain is enormous, the largest of any animal on earth—of any animal that has *ever* lived. Yes, I know, as I myself admit, I am a leviathan. Your brain is larger relative to your size. It is, what, two-and-a-half percent of your body mass?

And mine is far less. But there is at times something to be said for absolute size, don't you think? Absolute size marks intelligence in a way relative size does not. I understand, though, that you may remain stuck on relativistic notions, especially when they benefit you.

Just as your brain has evolved to suit the needs of a mid-sized terrestrial mammal within a complex social group living in a sometimes hostile environment, our brains have evolved to suit us. The oceans aren't often hostile (except, of course, when you have made them so). Our social groups are complex, but the situations we find ourselves in generally are not. We simply don't need some of the tricks of which you are so proud.

Your brain is better with olfactory information, and mine is better with auditory. Your environment requires a strong sense of smell, and mine does not. But directional hearing is critical for me. And echolocation—well, if you really understood it, you would be green as algae with envy. And geomagnetic locating...I won't get into that at all. Too much for you, I'm afraid. So we live with our differences: you find me smelly, and I find you noisy.

But both our brains are good at mental processing. We are both wired for intelligence, especially social intelligence. We each have individual personalities. We can make decisions quickly in complicated situations. We can take care of our families and make friends. We can have fun. We can provide help when necessary, and even sacrifice ourselves for others. We form cultures. We learn and pass on that learning to others.

So please don't make the mistake, the perhaps fatal error, of assuming that because you have a relatively large brain, indeed, a remarkable brain, that others lack intelligence. I have learned a lot in my roving. I dive deep, and I journey to and from the higher latitudes. I think—in some ways more clearly than you. And, most importantly for this song, I remember.

☙

Yes, we have cultures. Not a Cetacean culture. Not a Sperm Whale culture. But diverse cultures that we have passed on for generations upon generations. For thousands of generations. Tens of thousands. I'm not going to dive into any nature versus nurture argument here. I'm just not. But you need to understand that our societies are not only different from yours and Orcas' and Blues' but also from each others'.

My clan is spread throughout the Pacific, but our ancestral home is off the Galapagos Islands. Our Diaspora, especially following the massacres of my forefathers' time, is as far-flung as the Indian Ocean and the South Atlantic. And yet we remain culturally distinct. Our language is ours alone. No other clan communicates as we do. Our songs are never the same as others' songs. We tend toward greater synchronicity in our dances—even in our dives. We are roundabout, traveling more circuitous routes than other clans do. And yet we also venture closer to land which, fortunately and unfortunately, brings us more often in contact with you.

Though the ranges of our foraging inevitably overlap with those of other clans, we generally remain within our clan. Females, especially, live their whole lives with other females of their clan. Of course, intermingling occurs, particularly among roving males. Sometimes thousands of us, male and female from various clans, find ourselves together. Such moments can be celebratory, even raucous, but we remain cognizant of our distinctions. Some seasons we are more successful at foraging, and other seasons other clans are.

In our entire history, age upon age, no clan has ever fought with another clan. Such battles simply make no sense in the long sweep of time. Whatever our differences, we realize that we are present, always here and now, alive, bound together in this world, our home. And that understanding, embedded in each of our cultures because of who we are as beings, supersedes whatever differences exist. Our languages and customs may vary, but our connection to sea and sky, air and water, does not. It couldn't possibly.

⌒℁⌒

We sing and dance. Why? Because it's fun, but also because it brings us together. Sometimes we even learn a little about who we are and what we're all doing here. When I was young, my family and others in my clan convened near the surface most afternoons. We met in the best of times and the worst of times. We swam together, logged and rolled simultaneously, and hung vertically, our buoyancy pure. We danced belly to belly, jaws touching, sometimes clasping. We sang our codas, our patterned songs, as solos, duets, even choruses. We side-fluked, making our sharp turns together, and spyhopped to look at that amazing other world of air and sky. We breached and we lob-tailed, both of which are infectious. Invariably, once one of us shook the sky, the whole group went aerial. The more families that gathered, the wilder it got. It was all a cracking good time.

I was something of a lobtailing prodigy. I probably sound like I'm bragging again, but I was. Even at the age of three I could strike my flukes on the sea's surface with the best of them. When lobtailing, you can go vertical or horizontal—or really any angle in between—but I was an on-the-surface-straight-ahead horizontal lobtailing virtuoso. I raised my peduncle and flukes just so, and then smacked the water hard—the Crash and Splash. We all lobtail (the more the merrier, actually), but early on I developed a thwacking good Dorsal Down, the reverse lobtail. I'd gather speed, roll onto my back, dip and rise, and shake down the thunder. I also perfected the Cross-Over, a dozen quick ventrals and dorsals, back and forth for a minute or more with no break until I was exhausted and exhilarated.

Now, on those rare occasions when I am with the females and the young, I still enjoy singing and dancing—belly to belly and jaw to jaw. And I still like the attention that a snapping series of Cross-Overs brings.

I had a gift for lobtailing, but I admit that breaching is the most spectacular aerial move we make. There's nothing quite like launching yourself, hurtling airborne until the world's gravity grabs you and flings you onto the water. A relative calm with a bit of wind helps, but the real key to breaching is the dive. You fluke-up, dive fast and deep, flip, and shoot straight for sky. All out. Top speed. A twist in the air as you fly free provides a nice touch, but whatever you do, you have to give it your all—all in, all the way. The Crash and Splash can be beyond belief. A side-landing is best. Belly breaching tends to be jaw-jarring, and back breaching is tough on the blowhole and the spermaceti organ. Any time of day will do, any time at all, but a moonlight breach is especially gratifying.

Now, though, I rarely breach. Part of it is my size—fifty-nine tons is a lot of mass to hurl at the moon. And part of it is, I suppose, maturity. But part of it, too, may be that abandon, unabashed and unfettered, is harder to come by. I know too much about life and what is happening to the world.

I hope I'm not making my early childhood sound too idyllic. There were wonderful moments, to be sure. Some days and nights really were halcyon, the seasons turning with the currents and the temperature. We meandered the Eastern Pacific, foraging beneath the waves and congregating under the sun and stars. But beneath it all, an undercurrent of peril flowed through my family and my clan. It wasn't just the intermittent storms whose swells made me breach just to breathe—to breach, in fact, for my life. Or the occasional Orca attack that killed some distant aunt's newborn. Or the presence of sharks on the periphery of the clan waiting for some disaster so they too could strike.

No, the deepest threat was *you*. When I was young, you were still slaughtering upwards of thirty thousand of us a year. My clan survived my first year pretty much intact, but we were decimated my second. You might use words like *pod* or *school*, but when you do you miss the point. I may be a leviathan, but I am no fish. When I was young, I was part of a complex and sophisticated and, yes, *related* society that you massacred. Your huge steel ships slew and flayed us with impunity. And without so much as a thought that you were destroying an ancient and intricate culture.

We heard the Killing Fleet long before the ships loomed on the horizon. The deep roar of the Death Factory bellowed beneath the varying snarls of the Killers. We should have scattered in 360 directions at the first sound, but it is not in us to do so. As I said, we live together in a close, related society. And, so, we died together. We congregated in death.

The Killing Fleet moved inexorably toward us at eighteen knots. Most of us are faster, but outrunning them wasn't a choice because we move at the pace of our slowest member. And it would have proved useless, in any case. We tire, but the Killing Fleet does not. Their incessant din drove coherent thoughts from our minds. Panic raced through the clan.

In my forefathers' time, during the earlier slaughters, fighting back sometimes saved lives. The Killers were smaller, quieter, and less mechanical, the harpoons painful but not immediately lethal. Gnashing valiantly and fluking maniacally made survival possible. My famously white forefather destroyed your Killers, stove even your Death Factories, and sent you to the squids' domain. He survived all of your attacks to live a long life, wearing your bones as a talisman. And he was genuinely heroic. You killed his mates and his children and hunted him, and he avenged those he loved. Those he killed may have been brave, but they were not heroic. They died only so that someone far from danger could be a bit more comfortable.

In the killing fields of my youth, there was no place for heroism, only wanton carnage. Impregnable, utterly free from personal danger, your Killers fired harpoons that pierced our skin and exploded inside us. Blood smeared the sea. You murdered the children first, knowing their bereaved mothers would not leave. Confounded by your clamor and panicstricken by the slaughter, I clung to my mother's clicking. She and I dove, turned, went deeper, twisted, swam as long as we could, breached for a precious breath, and dove again. We found some distance from the insatiable Killers, but we couldn't escape the cacophony.

Only once was I close enough to watch the spiked jaws of your Death Factory hoisting the victims into air. Close enough to hear the horrific tearing apart of my kin, as though the sky itself was rent. Close enough to see my relatives' entrails dumped from clouds of steam into the sea for the sharks and other scavengers. The infernal grinding rose above the din. And though I have almost no olfactory sense, I smelled those boilers blotting the sky with their unholy steam. The stench lingers in my mind to this day.

My mother and I survived that killing season, just as my white forefather did before us. But, to hear my family tell it, he became bitter in the end. Alone. For him, never a good night or a gentle going. His rage against you fouled him from his scarred blowhole to his tattered flukes. His every song, even his last, remained a battle cry. I am not, as I said, heroic like him—but neither am I bitter. Or at least, I am trying not to be. My song is too important to be one of mere ire or past poison. No, the past is past. It is our future of which I sing.

We survivors went on living together—my mother, my remaining aunts and cousins, and I. My father was, of course, off in the higher latitudes, and my older brother had already left the family to rove with friends. But for the next year, my cousins, male and female, were all around me—not just in the afternoons when we played and sang, breached and rolled, lobtailed and spyhopped—but each and every day and night, drawn closer by loss. We never fought. Fighting only came later, with my friends and other males I didn't know. And that was over mating, which hadn't yet entered my mind.

My mother's voice always sang to me. Even when she dove, foraging for us, her voice carried to me. She provided me with milk the whole time. Though I was eating solid food by the time I turned two, she continued to nurse me—as much for the care as the sustenance. She touched me every day.

All of this ended abruptly for me when my mother's song went silent early in my fourth year. Ironically, it wasn't one of your mammoth ships, your Death Factory and your swift Killers, that murdered her. Tangled in your cable, she drowned.

❧

As my mother breathed before her last dive, the clouds broke. Sun streaked down, raining light around us and firing the water's surface. The storm had raged for three days. Lightning had webbed the sky, and the winds had spun white drifts across the water. Twenty-three of us had followed a deep fissure that cut from the abyssal trench almost to the shore—far farther toward land than we usually ventured, but the foraging had been good. All of us had been disconcerted by the ferocity of the storm, but the sudden light made me think danger had passed with the scudding clouds.

Flukes up, my mother disappeared. I stayed aloft, too young to dive as deep as she. Five of my young cousins were with me, all staying above our elders as they plummeted. The divers began to click in earnest after ten minutes, their songs spreading as they foraged. My mother headed deeper, as was her custom. Her voice rose to me as always. And she was successful, as she so often was. For forty minutes, I listened to her songs and rolled to watch the streaming light.

Almost out of breath, my mother began to rise. When her song first changed, there was little note of alarm. I may have been slow to react—I don't know because the flow of time had already begun to alter before I understood what was happening. My aunts and older cousins converged as though one of us had stranded. And, I suppose, in a way, my mother had. Her voice became strident. The more she struggled, the more tangled she became in your cable.

Her panic lasted only until she heard my stricken cries. She slowed her clicking as I dove. But I couldn't go deep at all. My breath kept catching in my throat. In my panic, I couldn't stay on the surface long enough to fill my lungs. I knew even then that I wasn't old enough to find her dark depth, and in that moment, helplessness washed over me. My inability to do anything struck me more deeply than any harpoon's point. But I still kept up with my shallow, useless dives. It was all I could do.

When her sisters and cousins reached her, my mother stopped struggling. Her song became softer, the love in her voice telling them—and me—to stay clear. And then as they sang to her, her song went silent. I don't think she was dead yet. She knew that if I reached her, I would die, too. And so she no longer called to me.

I dove on into the night. First my cousins and then my aunts, returning from the depths, stroked me. Voices clicked condolences around me. But I was inconsolable. The voice I needed to hear was mute.

I never looked for my mother's body. It was another year before I was capable of diving to that depth. She would by then have been only bone. I have never known, but I imagined that the deep sea sharks were already swarming before she let go her last breath. It is the way of the sea, the way of life.

Once her song was silent, she was gone. My family sang to me, but the voice I longed to hear was missing. There are no orphans among us, however. An aunt suckled me freely, as though I were her son. The nourishment kept me growing, but it never felt the same. I weaned myself in four months. When my family gathered in the afternoons, I lobtailed harder than ever. But, in truth, there was more anger than joy in the crack of my flukes against water.

I stayed another year within the clan, but early in my fifth year, I left with two older cousins who were heading west on their own. Was I too young? Probably, but I was large for my age and still torn from my family by grief. The Killing Fleet had brought us survivors together in our horror, but the nature of my mother's death, her disappearance into that dark depth, in some way isolated me. And so I went west, too.

☙

My cousins and I joined half a dozen other young males, but we didn't, at first, venture into the higher latitudes. Orcas and sharks were rare in the open ocean, and I learned to hunt Architeuthis and the lesser squid, earning my first scars out toward the North Pacific Gyre. At that time, your garbage had not yet gathered there.

Tsunamis passed us regularly, but they were no threat either. With my echolocation, I can often hear a distant undersea earthquake or volcanic eruption. The subsequent tsunami travels fast, sometimes at 450 knots, but it isn't dangerous. As the wave approaches, the ocean's level rises a few feet. The tsunami passes in twenty to thirty minutes, and the surface falls gently back to its normal level. If your hearing and echolocation aren't acute, you'd barely notice what's happening. Near shore, tsunamis slow to a mere one hundred knots, but the shape of the shallows can, at times, really focus the wave's energy—and all hell breaks lose against the land.

You loosed all hell of a different sort in the Pacific of my first youthful forays. The islands, Bikini and Enewetak, have exotic names and remote locations. Or, rather, *remote* for terrestrials like you, but not for my friends and me. At the far edge of our range, you detonated your most terrifying Marvels.

The first time, the increasing activity of your ships and airplanes let us know you were up to some mischief, but I was too young and naive to comprehend the extent of the horror —your astonishing knack for destruction. I guess there was no way for any sentient being to really understand what you were doing until you had already done it. How could I have been prepared? How could anyone?

The flash, though distant, inverted the day, reversed the horizon. The cloud rose like hell's own jellyfish. The heat slapped the air, then the water. The earth quaked, the sky erupted, and invisible waves of death poured over all of us living in the area. The drifting clouds and noxious rain came later, more slowly and more sickeningly.

Earthquakes, volcanic eruptions, and tsunamis are all natural events, occurrences in the life of the Earth—part of the planet's cycle. Yes, the seascape is sometimes altered. Some species benefit and others suffer. Your technological destruction is, however, unnatural. Your explosions, though small compared to the earth's strongest storms, are not part of any cycle. They rain death and obliterate life. In becoming the Destroyers, you set yourselves apart from the world and above the rest of us who live here.

But why? Why would anyone wreak this havoc upon the world? Even all of these years later, I can't figure out what possesses you.

⊂⊗⊃

Now, I am a loner. My song becomes one of solitude. The oldest and largest of us males have always been solitary. I don't mind it, really. Time alone is good. You can think things through, clear up what is in your mind, make connections among your thoughts, find the currents, the ebb and flow, the cold and the warm, and feel them purl.

After I left my family, I lived another seventeen years with friends, a fraternity of sorts. That's the way our culture works. But our fellowship eventually broke up over mating. That, too, is our way. And now, as I have said, I live alone. I never lived with my father. In fact, I only met him half a dozen times, twice when I was very young and he returned to mate, and four times, by chance, later in my rovings. I really only knew him through his songs. He was not bitter like my white fore-father, but his songs did have a motif, a recurring theme: The life of the mind is a lonely one. The price each of us pays for thought is solitude. He was, I think, right.

But I do still, after all of these years, miss the others. I will at some point in this journey visit the females and the young of my clan. I won't mate, though. I will not bring any more young into this world. But I miss the attention from the females when I mated. I miss the cavorting with friends and family. I miss the interplay, the playfulness. I still need on occasion the company, the companionship, the touch, the feel of others around me. I remain, even in my solitude, irresistibly drawn to others.

Indeed, even to you. For the most part, you no longer mas-sacre us. The Killing Fleets dwindled, eventually eradicating themselves even as they attempted to annihilate us. And then the Killers vanished. There are rumors that they still hunt and hew near the Asian islands, but I haven't myself heard their ominous rumble in decades. Recently, you have, on occasion, even tried to save us.

I don't want to offend you because I need you to continue to listen to my song. But you have to understand that you are still destroying us, all of us in this world. Despite all of the Marvels you have created—the artifices and the ships and planes that move you so swiftly through the water and air—you have forgotten so much of what you need to know to live, to keep this planet, our home, alive. Your mechanical Marvels become ever bigger and faster, but your life does not become better. Not really.

We have no Marvels because we need no marvels. We live in the sea as part of the sea. To justify your excesses and your destruction, you have in the past demonized us, made us symbols of evil. Now, for the most part, you simply mythologize us. But we are living, breathing, thinking...need I go on? You have gone on producing your Marvels—and missing the marvels all about you in every moment.

I love the horizon, despite the Killing Fleet that loomed there in my youth. During my roving, I often spyhop, looking out at this juncture of worlds, sometimes serrated, sometimes a fine line, and sometimes a nebulous merging of sea and sky. I like that this boundary between my domain and the domain of the birds can be at once exact (this is *air*, and this is *water*) and unclear—always the same and always changing. And I admire the birds, their flight, their soaring and swooping. Their swimming through air.

When, as now, I spy an albatross, I know I am arriving in warmer waters. I am venturing far enough south that I will find companionship soon enough. A solitary albatross is for me, paradoxically, a harbinger of clan and kinship. And yet, like me, the albatross is always wandering. Large for a bird, with a wingspan of ten feet or more. Tenacious, able to cross vast stretches, thousands of miles, of open sea. Never rushing and never resting. And never returning home except to breed.

I suppose I see myself in the albatross, too. A kinship based on neither clan nor genetics. Always roaming. Always searching. Always looking off toward the horizon. Sometimes, I don't know how, I seem to see beyond the horizon. Can the albatross? I don't really know. But perhaps you could, if you let yourself.

❦

Your din has become more than an annoyance. Even out here in, as you would say, *the middle of nowhere*. Our world, you must remember, is essentially acoustic, and you have made the sea increasingly noisy in my lifetime. Your tankers become ever bigger, faster, more numerous—and louder. Sometimes when they roar by, I cannot think.

Our lives are anchored in our listening; it is what we do. Our societies are formed on communication, on our need to sing and to hear one another clearly. Unnatural low frequency sounds, the sort you make, disrupt our culture. You bombard us with seismic pulses. Sonar is worse, cleaving families and clans. No one, male or female, feels much like mating amid your incessant cacophony. It affects our diving and, therefore, our eating; it sometimes causes stranding.

We avoid your clamor when possible. But it is becoming less possible. I can venture farther away, but it isn't so simple for families, much less clans like the one I will soon visit. An extended family of hundreds of females and offspring can't simply leave fertile foraging grounds. They have to put up with your noise, and damage to their hearing is inevitable. But it isn't just our hearing; noise tears the very fabric of our lives.

Does this sound at all familiar?

I admit (in fact, I should have already) that I spend a lot of my time alone because of females. Or, to be more candid, because of the way we males are when we are around them. To put it bluntly, they outperform us. When we are together, they are more successful at foraging—about twice as successful. Part of that may, of course, be that when males and females are together, males don't always focus on food. Females pay better attention to us, to each other, to our young—and find more food.

But that's not even all of it. The older and larger I've become, the more I realize that females' outperformance of us is relative and absolute. And my mind certainly is not on mating the way it once was. My clicks these days are no longer the slow rhythmic sounds that herald a male's needs. No, my clicking is faster; my needs now are at once more catholic and more urgent.

Females are simply better in some ways than we males are. We grow bigger—in fact, we are the most sexually dimorphic of mammals—but our size doesn't make us better. Bigger, you see, isn't necessarily better.

I do not really want to bring up mass strandings, but I must. It is something we do. My reluctance isn't based on squeamishness or on the fact that my brother died in a mass stranding. I was not, thankfully, around when he stranded with eleven of his friends on the coast of New Zealand. I was only thirteen at the time, six years younger than he, and I probably would have done something rash.

The manner of death is slow, painful, and humiliating. You lie there wholly out of your element on that gritty shore. You gasp. Your chest heaves as gravity crushes you. You slowly overheat, your own weight making you hemorrhage. You slap your flukes, causing the earth itself to shudder, but you can do nothing to move yourself back into water. You blink and roll your eyes as you suffocate. But that is not the worst of it. No, the real horror is what drove you onto that alien shore in the first place.

We strand out of altruism. It makes no sense, but there it is. Yes, weather and ocean conditions, exacerbated by your noise and pollution, contribute to any initial stranding, but the truth that's so hard for me is that mass strandings are essentially social. When one of us strands, others follow—out of a necessity that's hard to understand, much less explain. My brother was not the first ashore that day in New Zealand.

I have only been present once for a stranding. I was forty-nine, returning to the Gulf of Alaska from a visit to my clan in the lower latitudes. My foraging for cephalopods led me in toward the coast of the continent. A truly violent storm, one that reminded me of the day my mother died, had just blown ashore. Clouds roiled, and lightning clawed the water.

I heard the alarm clicking across a great distance. The coda was not my clan's, but I still ventured shoreward, unsure of what was happening. Or, rather, sure of what was occurring but unsure of what I was doing. When I arrived at the long tapering cove, the storm had vanished over the cliffs above the beach. Its only vestiges were the eight young males glistening darkly in the sun. Light spilled across the rolling surf, their gray-brown bodies, and the pale jutting rocks. You were already infesting the area, your machines whining.

I stayed offshore, but I was, I admit, disoriented. None of the stranded eight was my kin; their dialect was unfamiliar. All were less than half my age. Though I was mature, level-headed, seemingly detached, something still pulled me toward shore. Something inside me still called me to follow. I could not save a single one of them, nor could I leave that cove until the last one expired two days later. I swam uselessly back and forth beneath the thwacking of the whirlybirds. Sirens intermittently pierced the air; machines ground louder than the waves. I forgot to eat, forgot, in fact, everything except the stranded young males' impotence and my mother's vulnerability and my aunts' perplexity when they went to her but could do nothing to untangle the cable strangling her.

Part of my consternation came from our inherent need to follow, but part went far deeper. That social cohesion that, as I have said, I still miss after decades of solitary roving bids us all to convene, even in death. Just as we did with the Killing Fleet. We are, as a species, in life together. Our willingness, our *need*, to help each other overrides everything else. The *curse* is stranding, but the *blessing* of belonging, genetically *and* culturally, outweighs it. It's part, the essential part, of being alive.

When I'm close, I can hear the clan's clicking. It's early in the day, and the females and their young are spread out over dozens of miles foraging. As I approach, I clang my arrival every five seconds or so. The first to see me are a mother, her two-year-old, and an older aunt. Though they don't know me—only the aunt has ever met me—they are unabashed in their greeting. All three rub me as we dance along, my old skin sloughing. I roll gently, basking in their movements, my role to graciously receive their ministrations. My clangs quiet to clicks.

I've already heard two other large males in the area, but they are bent on breeding—and I am not. Their clicks are slower and deeper than mine. I will tolerate their presence; they will be less tolerant of each other, a loose alliance that will soon break. The females' relationships are far more complex, an interwoven pattern of personal preferences that are most often lost on us males.

Soon, the matriarch appears. She and I mated long ago, and our son has been out roving for more than twenty years. She looks older now, her dorsal calluses thicker and her flukes more worn. Her codas, though, are tender as her jaw grazes affectionately along mine.

In the afternoon, the clan congregates. Or, rather, *we* convene, thirty-four of us, for I remain the center of attention. The newborns especially are drawn to me, my size making them giddy as they slide along my belly and back. Few of the young have met me before, and many have never even heard of me, but they gather around me as though I am some peripatetic grandfather home at last to spin his seafaring tales. And, I suppose, that's exactly what I am.

The energy pulsing through the water intoxicates me, and before I know it I'm loosing a splendid series of lobtails, Cross-Overs that shake the sky. I am, for a moment, young again, touched by others and untouched by Killing Fleets and tangled cable and lethal Marvels and Black Death and monumental garbage and mass strandings. My joy is contagious. The young breach with ecstasy, crash-splashing until the water froths. Their mothers lobtail their ovations, and even the old aunts and the matriarch side-fluke together, turning and turning in our ancient dance. And I…I spyhop to take it all in.

I have been alone so much for so long that I have all but forgotten…No, that's not true. I have never forgotten that when I was young I was touched each day, each hour. My mother's touch mattered to me as much as her milk. Her nudges and strokes, her nuzzling and rubbing, meant the world to me. Whenever she returned from a deep dive, I swam her belly before feeding. She ran her head from my jaw to my flukes. My skin, already wrinkled even then, tingled every time. When my mother died, my aunts took over. It was not the same, but it was still good, still necessary. And my rolling and sliding with my cousins cleaned us—and often provided the best moments of my day. All manner of touch matters. Perhaps not as overtly as hearing or seeing, but in its own way just as importantly. And it's good to be back now, even for a few moments, among those who touch me.

❧

The clan is good, all in all, though we've never quite recovered from the Killing Fleet massacres. Our numbers are not what they once were, and our birthrate is down. Not yet inexorably, but dangerously. We remain close to the edge of that deep trench into eternal darkness. The threats now are more subtle, but every bit as real—and far more pervasive. The sudden swooping of the Killing Fleet has been supplanted by the gradual menace that I feel in this current.

As we approach the winter solstice, the water here in the Pacific near the Galapagos is tepid—warmer than when I was young. The increase isn't yet great, just two or three degrees, but it may make all the difference in the world. Warm water is pooling here, and the earth's cycle is being altered. Warm water buries cool water. Cold, salty water sinks, taking nutrients with it. The thermocline, which used to be about 140 feet here, is now about 500 feet. Most of the fish and small squid are gone. The clan has to range farther and dive deeper for food. But this isn't about us. Not really.

It's less windy now, except during storms, which grow more fierce. The air is both warmer and wetter. Ten of the last fifteen years have been the hottest I've known. Storms are more frequent as well as more intense. Is this a fluctuation in my lifetime, or something permanent? Is it merely a difference of degree, or a difference of kind? When does a difference of degree *cause* a difference of kind? I don't know, but I do know that something is happening here. Something that's trying to teach us how small our world really is…how interrelated all of our weather is—from my summer grounds in the Arctic to these equatorial waters in winter…how bound we all are in this moment….

Could I be wrong about this? No. These climate changes are not merely the flow in the ebb of one individual's life. They are occurring. My clan *lives* amid them. The only question is, How drastic will the consequences be?

⚭

The sudden silence tells me something is wrong. At the edge of the clan's range, I am at depth, foraging in total darkness, when the clicking above me stops. Orcas. Our only predators other than you.

Killer whales is a misnomer—Orcas are killers, but they are no whales. They are *blackfish*, closer to dolphins but incorrectly perceived as one of us because of their size. Though I don't, you would think them huge—typically, twenty-four feet and nine tons. They don't really look like me either. Black and white always, with white patches behind their dark eyes.

A rounded head that tapers to a point, and a prominent, even pointy, dorsal fin.

Though they are our natural enemies, I must admit, grudgingly, that they are fast and smart. Their top speed, thirty knots, is one-and-a-half times mine. Their brains are larger, relatively, than ours, though not absolutely. They remain in close-knit families their whole lives, each clan singing in its own dialect. And, they always work together, especially in their ferocious attacks. If they did not murder our young, I would tolerate, even respect them.

I stop clicking and start rising, the continued quiet telling me that the attack hasn't yet begun. I break the thermocline, pass into warm water. By the time I reach light, mad clicking showers me. The sorties have begun. The family under attack, three adult females, two immature males, and a baby, has taken to the surface, a smart move. There is no refuge in the open sea, and no hope of running. The surface takes a dimension from the enemy (at least, an *aquatic* enemy)—an attack can't come from above. And, you can catch a breath without breaking ranks. Unfortunately, the family has taken a flukes-out defensive position. When we're attacked, we form tight circles, our young in the center, our heads or our flukes facing the enemy. There are advantages to each—we gnash *and* we flail—but I favor heads-out so you can see exactly what's attacking you.

Eight Orcas, including two large males, are flashing at the family's formation, trying to breach it. Blood is already in the water.

Rage. My power still surges in moments of ire. Water flees, and air shivers. The Orcas strike repeatedly at the smaller mother, driving her from the defensive circle. They are not after her —they know her baby will panic, and the formation will disintegrate. White light exploding in my brain, I hurl myself at the three Orcas slicing her from her family.

Two of the Orcas fly at me. One arcs at my back and bites hard just aft of my head crease. He shakes his head to tear me, but my tough skin and thick blubber slow him for a second —a lethal second. They want blood in the water, they will have blood in the water. All in, all the way, I roll and rip my flukes, crushing his pointed snout and mouth still agape with a chunk of my flesh. Turning his brain to a sponge. I spin as the second Orca, a larger male with a wavy dorsal fin, slashes me across the head, his teeth raking me from blowhole to upper jaw. I jerk my head, my jaws clamping him just fore of his flukes. If I had upper teeth, I'd cut clear through. I soar in a world of silver light. Blood pounds in my head, swirls around me in the water.

Still clenching, I breach and whip the Orca, all nine tons of him, against the surface. Stunned, he starts to roll. I am at his neck, clutching fast, ripping out his throat. Blowing. Sucking air. Submerging. Torpedoing at the third Orca the wounded mother is fighting. Seizing its whole head. Mauling. Mangling. Light flaming all about me.

The final five Orcas run from my fury. As I escort the mother back to her family, I am exhausted. Utterly spent. The world fades to blue again—blue and red. The blood running from my snout is warm. The salt water sears my injured back. The mother is slashed below her right flipper; tentacles of torn intestine trail her. I will live, but she will not.

The large Orca with the wavy fin sinks slowly, his white belly up and his bright blood pluming. The first Orca slips sideways, the hunk of my back still in his mashed mouth. The third's pool of blood spreads, a signal for scavengers. Mother and baby nuzzle and nudge. The family rolls around me, clicking approbation. But I only swim away. I must be alone—alone with the savage truth I have always known about myself. I am the *killer whale*.

When I've put some distance between the family and myself,

I surface. Spyhopping, I see nothing but sea and sky and high sparse drifting white clouds. The swell is gentle, the wind light. Blood still seeps from my head and back, but the physical wounds aren't my concern. My blood may well attract sharks, but even they are not stupid enough to attack me in this moment. My anger slowly evaporates in the vastness of the open ocean.

I'm not sorry. Not at all. I was provoked, and I responded. But I am frightened by my fury. I don't otherwise harbor fears. The world exists. Death occurs to each of us no matter what currents we have swum. And pain is pain. It occurs, too. It is part of life, part of the great current. I am, as I have said, a gentle soul. I love touch, but I'm too shy, too solitary, to stick around for the family's adulation. I was not heroic—I was incensed, and the white light of rage flowed through me.

Can the world itself experience that infuriating flash of light? You wouldn't think so, but the world's time is far deeper than our time. The Earth breathes just as we do—but only once each year, a single inhalation and exhalation. We find it difficult to fathom this breadth and depth of time, the world's slow cycle of life. But what is happening here in the Eastern Pacific and there at the North Pole—the melting and the calving, the pooling and the storms—is a mere instant in the world's life. At what point does the destruction and pollution, the emissions and toxic wastes—the flotsam and jetsam that you insist on flinging ever overboard—provoke the planet? Is this the moment, the blind blink of deep time?

⚬∞⚬

In the aftermath of the Orca attack, I swim westward, farther into the ocean's expanse, away from the clan to which I belong. The horizon retreats in every direction. I need this time alone. I know all too well that we exist in time, but often it flows unevenly. It fluctuates. Now, for instance, time is an eel. It slithers and twists, turns, stops, waits, strikes. I keep on my way until I can find again the pulse of this planet, the regular beat of the world, and finally rid myself of this eel that stalks me.

The life of the world is not eternal. It is so vast that it seems eternal, but it is not. Each of us is alive, a part of this living planet, for only a short time. But the eternal does exist—and you can discover it during your brief time. It is here and now. It exists for me in the suspended moment, in the balance between rising and falling as I breach. In the moment when Architeuthis and I strike at once. The moment in which the newborn, nudged to the surface, breaks into sky and takes his first breath. The moment in which, while I am spyhopping, the sea stops and the light unfurls.

Despite the tone of my voice, the movements of this song, and the refrains to which I return, I do not hate you. As you have heard, I'm perfectly capable of fury, but I am not sunk in animosity. I cannot sustain malevolence. I understand that you have become so enamored of your Marvels that you are blind to their consequences, so bewitched by your creations that you ignore the concomitant destruction. But it is time, your moment in time, your moment beyond time. You must discover the world anew and let it take your breath afresh—and you must fall in love with it, all in, all the way.

Just as I find it hard to hate, I find it difficult to love. I loved my mother, was devoted to her, would have defended her, tooth and fluke, until the end. I never knew my father well enough to love him. The idea of him, yes, but not him. My siblings? Cousins and friends? Yes, I've felt loyalty and caring. I would have, in certain situations, given my life for them. The females with whom I mated? In those times, perhaps. And later, when they were the caretakers of my offspring. My children? Well, yes, them, too. But I never lived with them, never was the parent each of their mothers was. And I always moved on, driven both by our norms and by something inside me.

But I am able to love this world—and life itself. A deep and abiding love. And that, *sometimes,* is enough. Not always. But sometimes. You are built for love, as I am. As are we all. This I know: I love the sea and the air, the seasons, the journey that is my life, the life of each and every one of us, each day and every night. I love each blow from my spout when I rise from a long, dark dive. Each and every light-blasted breath.

And this is why, against all odds, I offer you this song. So you, too, may love every moment of every day. Each breath. Each moment in and out of time. The spreading light and the gathering darkness. Warm air and cold rain. Every morsel of food. The presence of others. Voices singing. The touch of this being and that. Even pain, great and small. All of it.

Pressure mounts. The water is warm, above 80 degrees, and the air thick. Clouds tower, but there is no front. Heat and vapor rise. Thunderstorms converge, and wind soars. The bands of clouds are already radiating. Something violent this way comes, and the gathering storm will tear the stillness, tear every thought we have.

This terrible beauty beginning to whirl around me makes me think about the mystery at the core of each of our lives. The mystery from which all things—sea salt and sharks' teeth and ice songs and storm centers—emanate. Life is, in essence, aqua incognita. But this mystery is just that; it cannot *save* the world now. The current changes suggest that you have placed your Marvels above our mystery. Or worse, that you have set yourselves up as Destroyers, bringing devastation down on all of us. You need humility—before the sacredness of being, the interwoven fabric of life in the sea and sky and on land. You must again remember, come once more to understand, that you are part of the fabric of life, not separate from it—and certainly not above it.

⌗

The storm is immensely powerful, the most potent event on the planet. Your most destructive Marvels are minute by comparison. The seas grow so rough that even the most titanic of your tankers flee the region. The wind and the electrical noise drown every sound you make. The swells rise to forty and then fifty feet, the periods between them brief and erratic. Spindrift screams, and clouds streak at over 120 knots. On the surface, what is air and what is water becomes moot. I must breach, as full out as when I was young, to catch my breath.

I go deep to avoid the chaos in the world above. The power of this sea, the energy in this storm, is enough to make me fathom unknown waters, and enough, in and of itself, to make me understand nothing. Nothing but the existence of the world. Nothing but this storm. The universe shrinks to this black hole of energy, this storm, here and now. The Earth in this moment becomes the mystery unadorned.

⚬❦⚬

I breach in the eye of the storm. The eye is small, perhaps twelve miles across. Clouds in the eyewall rip by at 150 knots, but here the wind is only a tenth that speed. Albatrosses, trapped by the wall's dark clouds rising to fantastic heights, will go wherever the storm takes them. The storm itself is traveling toward higher latitudes at about ten knots, and I can keep up for a time, swimming here in the relative tranquility of the eye. The seas remain rough and confused, but I am not. All about me, energy whirls—perhaps a million cubic miles of atmosphere is enclosed by the storm—and I exist for a moment in the still point at its center. Energy flows through me from the wounds in my head to the torn tips of my flukes.

It doesn't matter what you call that which binds us, each to the other and all to the sea and sky and land. In this relative stillness, I can feel the flow of the world. And when I am roving, I can as well. Not merely in the swells or the current or the thermocline. It is something else, something more, something that permeates the world but isn't evident unless you yield to it. Wherever you are, whatever you're doing. I've thought about all of this, and I understand that I can't know it in the way I know currents and depths and the pull of the poles. I also realize I must sound like some quixotic rhapsodist breaching beneath a full moon. But this energy abides. I—we all—breathe it each moment. And it is, ultimately, why I harbor no bitterness.

⚬୫୦

The vast power of this storm reminds us all that our world is, indeed, small. The storm will affect weather across the planet. It will touch every one of us in some way. It is not merely the albatrosses that will be displaced. We will all, in ways large and small, be disheveled. This is no mere cod flapping its tail: currents will be altered and winds disturbed. Food resources will decline, destroyed or redistributed so that many will starve and a few will grow fat. It's the way the world works.

Though the storm cuts a wide swath, I swim against its prevailing path, breaching intermittently for breath, and thus in three days free myself of it (if not its consequences). The Orca's slashes across my head are scarring, but my dorsal wound festers. Perhaps, ironically, I am being eaten by bacteria.

I'm also, though I'd rather not admit it, a little worn down by the Orca attack and the storm. Don't misunderstand me. I'm not at all tired of this song. But I need to find seas less vexed and winds more tempered. I am feeling, well, a little old—and yet I am in no way ready to feed the sharks. I still must seek a newer world. I must continue to strive and search and sing. As must we all.

⁂

The vestiges I find in the storm's wake disturb me. Remnants not just of its power but also of your pollution. Dead albatrosses and gannets are scattered about, of course. There is always carrion for scavengers after a potent storm. But I also encounter small floating islands of debris—not just torn branches and tangled vines, but your synthetic refuse blown into the open ocean by this storm. Your detritus is indigestible, but that doesn't stop marine life, even Cetaceans, from ingesting it. We eat what we find, even when it's inedible.

What I can't see is even worse than what I can. Chemicals and other artificial particles I can barely discern taint the water. Storms invariably stir the toxins you have discharged into the sea, and plastics, it turns out, are pelagic, too. Perhaps 50,000 adulterated motes float in every square mile of ocean. And, rather than decompose, they disintegrate into smaller, but no less noxious, specks.

You have, I understand, used the sea as a garbage dump for a thousand generations. But in my lifetime, the life of one being, the problem has become exponentially worse—so much so that this storm's detritus is a mere stain compared to the debris already heaped and swirling within the North Pacific Gyre. The mass of plastic floating there is five or six times that of the zooplankton. In fact, the trash cycling within the gyre is so thick that no Cetacean, Baleen or Toothed, would traverse the area anymore. And it was, you must remember, not really so long ago that my friends and I wandered there unbound and unbounded. Now it is an unnatural archipelago well on its way to forming a septic continent mid-ocean.

Or worse, the garbage patch is becoming the malevolent eye of a far slower but far deadlier cyclonic event, a storm of sewage you build each day of every year. The ocean is one entity, one organ—and when you contaminate one part of it you inevitably foul all of it. You are currently disgorging your waste so fast that none of us can fully comprehend its impact on life. Not just marine life, but *all* life.

These dark thoughts drive me north and west, farther from my clan, farther from decomposing Orcas, and farther from the storm's wake. In fact, beyond Bikini and Enewetak, those infamous islands of my youth. I could keep going. Tempt fate. Venture closer to the islands where your Killing Fleets still loom and the Death Factories still operate. I could *attack* their Killers—I am capable of it—but it's a futile gesture. A suicidal mission.

I have always understood, of course, that you are the ones who operated the Death Factory, disemboweling and tearing apart my kin. You detonated the heinous Marvel and loosed the Black Death and laid the strangling cable. You persist in defiling the sea with your excrement. I am the Killer Whale, and you are the Destroyers. We must understand that about each other.

And, we need to go on. It's time. You have become the Destroyers, but you remain the world's best and only hope. I cling to this belief as I once clung to my mother's nipple. We are the only ones able to sing this song, and you are the only ones capable of undoing the damage you have done. I know that the very egocentrism that drives so much of your lunacy may now compel you to rescue our world. Here, after all, is our final paradox: in the face of the current climatic cataclysm, self-interest and altruism become one. The only way to save yourself is to save us all. We, as I have been chanting all along, are, and have always been, inexorably entwined—bound fast to each other and to this world. Now is the time to take action, to take your place as we wheel through our lives and the life of this planet. This is your brief moment. Seize it.

⚭

No, I will not venture toward the Islands of the Killing Fleets. I understand that it is better to go on singing my song than to die far from my clan. I am pelagic, even peripatetic, but my song belongs to all of us.

It is my role, given my age, my gender, and my size, to wander alone, to recall moments in my life, and to sing for you. My song matters to our survival, but the song is meaningless if it were not for others, us and you. We are all in this together as living creatures, as mammals, as social beings who share a home.

I breach under a gibbous moon. The vastness of the night sky makes me feel small. And the sheer number of wheeling stars mid-ocean takes my breath. The Milky Way spreads light like phosphorescent krill across the sky. A shooting star flares —an ephemeral traveler, as we all are.

I lobtail, a fine Dorsal Down that cracks the heavens. I become drunk for a moment on air and memory. I smack a dozen Cross-Overs. Though the festering wound on my back breaks open, I am as giddy as a two-year-old.

I suck air as though it is life. It *is* life, you know. We *are* air and water. Sometimes I am acutely aware of it in every cell—as though I were just born. The night air is bright, inside me and out—and the dark sky shimmers. Energy swirls around me, cycling through me and this world. Water flows in unfathomable currents. The air clicks and claps.

JAY AMBERG is the author
of twelve books.
You can contact him at
jayamberg.com.